# Evolution:
# A Guide for the Perplexed

Dwight F. Mix

Professor Emeritus

University of Arkansas

# Contents

# Fact or Theory?

Facts are stubborn things.

Our second president, John Adams, made this statement during his defense of the British soldiers who were accused of murder in the Boston Massacre. Despite his hatred of British rule, John Adams successfully defended the soldiers with his vigorous defense. The harshest penalty handed out was that the commanding officer had his thumb branded.

Evolution is a fact. Every creature living today arrived here by evolution. We cannot change facts, they are stubborn things.

Evolution is also a theory. How can it be both fact and theory?

In his revealing book, "Finding Darwin's God," Kenneth R. Miller explains that the word *evolution* is used in two different ways with two different meanings. He calls one meaning "history" and the other meaning "mechanism." He explains it this way:

## History

In 1794 William Smith, a land surveyor and map-maker, was responsible for choosing the route of the Somerset Canal. This canal was dug in Southwest England to transport coal by barge, rather than by pack mule and wagon.

Smith wanted to avoid porous rock in the lining of his canal to keep water from leaking out. He knew that certain formations in the upper layers of sediment indicated porous rock beneath, so he needed to identify the various layers.

These layers of sediment are laid down over millions of years. If you drive down any modern highway where the road bed is cut through a hillside you can see these layers. For a moderately sized hill you might be looking at 50 million years of the earth's history.

Smith soon learned that he could identify the layers by the fossils they contained. He found that the fossils in the top layer were identical to present-day creatures, and the ones just below were at least similar if not identical. As he went down from one layer to the next he found gradual change in species. In the bottom layers he found fossils of creatures that he could not identify with any living species. What he observed was change in species over time. That's evolution!

Evolution is change in species over time. That's what it is, that's all it is, simply change over time.

Smith may or may not have been the first to notice this, but many others have observed the same thing since. You can too. All you need is a steam shovel and a place to dig.

In addition to the fossil record, there is other evidence for evolution. Of course, evolution occurs over a long time span compared to our lifetime, making it difficult to "see" evolution. Nevertheless, it can be observed directly. One example is the AIDS virus. There was no such virus before about 1930. This virus evolved from other strains of virus and spread rapidly until people became aware of it and its properties.

Another example is purposeful breeding of livestock and race horses. The changes are small during our lifetime, but they indicate that change does occur.

The Russian geneticist Dmitry Belyaev (1917–1985) bred silver foxes for tameness. By breeding the tamest foxes from each generation he succeeded in producing animals that looked and acted like border collies, all in a matter of 20 years.

3

For a detailed look at this type of evidence I recommend "The Beak of the Finch," by Jonathan Weiner. It is the story of observations made over several years of the wildlife on the Galapagos Islands. These islands make an ideal laboratory because they are isolated. Any changes that occur are influenced only by the natural selection that occurs within the group of animals on the islands.

The climate varies in these islands from year to year. Some years there is plenty of rain, while in others there is drought. This leads to changes in the animals from year to year that are recorded in this book, thus illustrating the type of gradual change that results in evolution over longer periods of time.

Finally, another evidence for evolution is the tree of life. There is a clear order in the living world. The diversity of life is highly structured. The pattern of groups within groups is obvious.

Humans are most similar to chimps. Humans and chimps are similar to gorillas; humans, chimps, and gorillas are similar to orangutans, monkeys; and lemurs. This makes up the bulk of the primate family.

The primate family belongs to the larger group of mammals; mammals belong to the larger group of vertebrates. This tree of life continues with vertebrates combining with other animals to form the animal kingdom, which in turn, combines with the plant kingdom, and all this combines with other life forms to make up all living things.

The explanation for this is common ancestry. That is, humans have a common ancestry with chimps; humans and chimps have a common ancestry with the other apes, and so on.

Darwin thought that mankind originated in Africa because of this similarity, but the prejudice against such an idea was so widespread among scientists and non-scientists alike that they refused to even consider the idea. They looked elsewhere until Louis Leakey (1903–1972) showed the world that Africa was, indeed, the birthplace of humanity.

## Fossils

A fossil is an object that reveals information about the past. Fossilized bones are special because they are the most common fossil, and because they last a long time.

When an animal dies everything deteriorates. The soft tissue goes first, followed by the ligaments, then the bones, and finally the teeth. After an extended period of time nothing is left to indicate that the animal ever lived, except in special cases.

If the dying animal falls into a bog or marsh, minerals seep into the bones before they deteriorate. This does not stop the organic material (the original bone) from deteriorating, but it leaves the mineral deposit in the same shape as the bone. In other words, it is a rock, and rocks last a long time. Fossilized bones of dinosaurs are between 65 million and 235 million years old, yet they remain intact for us to find them.

## Mechanism

The other meaning of evolution is the theory that explains how evolution works, or Kenneth R. Miller's mechanism. Some theories explain a fact, while others are mere speculation. The theory of evolution explains a fact.

Most people don't understand the theory; they think it works like the following fairy tale:

"Isn't he cute," said Brown Bear.

Now mamma, you know he looks odd," said Papa Bear.

"Well, he doesn't look like you or me," said Brown Bear, "but you have to admit he is cute and cuddly."

"Let's name him Teddy Bear after our President," said Papa Bear. "Teddy Roosevelt is cute and cuddly."

"OK, since we just gave birth to a new species we should give him a new name."

What's wrong with this story? A lot! Bears don't talk, and Teddy Bears are not real. But the most glaring error for a scientist is the sudden appearance of a new species. Evolution does not work that way. No one has ever seen a new species produced. Instead, evolution works gradually over many generations. New species certainly do appear, but it takes a long time.

Skeptics are quite correct to say that a frog and a snake have never produced a lizard. But that does not mean that evolution is wrong. It just means that evolution does not work the way many people think it does. It works the way Wallace and Darwin said it does.

## How Science Works

Science is based on three key principles:

1. Follow the evidence.
2. Theories must be falsifiable. This means that you must be able to test a theory, and it further means that a theory is false if experiment does not agree.
3. Truth is derived from experiment, not from the comfort of speculation.

Science observes the world and tries to understand it. But this is difficult. Nature is complex, in fact so complex that any explanation of how nature works can very well be wrong. Since it can be wrong it cannot be called a fact. Instead, it is called a theory.

The magic words here are *how nature works*. If it explains how nature works, it is a theory. (There are other types of theories, so an explanation of how nature works is only one type.)

Once a theory has been proposed it is poked and prodded and examined to see if it holds water. It usually takes some vetting to get it right. Most new theories have some bugs if they are not totally wrong.

It is easy to prove that a theory is wrong. All it takes is one experimental result that contradicts the theory.

On the other hand, it is almost impossible to prove that a theory is true. No matter how many experiments agree with the theory, the one that disagrees could be just around the corner.

## Natural Selection

Darwin's theory is the culmination of several attempts by others to explain how evolution works, including Darwin's own grandfather, Erasmus Darwin. The best known of these earlier attempts is due to Jean-Baptiste Lamarck (1744–1829).

Lamarck's "transmutation theory" proposed that spontaneous generation continually produced simple life forms that developed greater complexity in succeeding generations.

In the 18th century there was widespread belief in spontaneous generation of life. One recipe for the spontaneous generation of mice was to pile old quilts in a corner, sprinkle generously with corn, and soon you would have baby mice.

In Lamarck's theory, traits could be passed on to the next generation if that trait was used by the parent. For example, giraffes have long necks because their ancestors reached for leaves high up in the trees.

Charles Darwin's theory became known when he was forced to reveal it by Alfred Russell Wallace (1823–1913). Darwin (1809–1882) intended to publish it after his death because he knew it would be controversial, but Wallace forced his hand when he sent Darwin a paper describing his (Wallace's) theory of evolution, which was the same as Darwin's. In his letter he asked Darwin to publish his paper if he found It worthwhile.

On June 18, 1858 Darwin received Wallace's paper describing natural selection. Wallace was doing field work in Southeast Asia and Darwin was in the process of writing his book. However, Darwin's 2 year old son was gravely ill with Scarlet Fever, so he was in no condition to deal with such a crisis. He sent Wallace's correspondence to his scientific friends, Joe Hooker and Charles Lyle, and ask them to deal with it.

Darwin had revealed his theory to Hooker and Lyle through his correspondence over some 20 years. There was no doubt about his priority, but Wallace's paper also deserved credit. What to do?

Hooker and Lyle read abstracts from both Darwin's and Wallace's work at the July 1 1858 meeting of the Linnean Society. This established Darwin's priority and gave Wallace the credit he deserved. Darwin did not attend because he was grieving over the loss of his two-year old son, who had died a few days earlier.

Darwin immediately began work on his "abstract" titled "On the Origin of Species by Means of Natural Selection, or the Preservations of Favored Races in the Struggle for Life." It was published on November 22, 1859, and was an unexpected success. The entire stock of 1,250 copies sold out immediately.

Evolution is simple, really. Two facts form the theoretical basis that explains evolution. Given these facts and enough time, change in species is bound to occur.

**Fact 1: There is variation among offspring.**

If you are a parent with more than one child you know about variation among offspring. If there is some variation among siblings then there is even more variation among unrelated offspring.

**Fact 2: Organisms produce more offspring than can possibly survive to reproduce.**

This fact is obvious among plants, insects, and smaller animals because each set of parents produce many offspring that never survive. Think about all those whirlybird maple seeds that clog your gutters in the spring, or the large number of seed ticks produced by each mother tick. Most of these never reproduce. This fact is less evident for large animals because they don't produce as many offspring. It was painfully evident among humans before the advent of vaccines and the discovery of penicillin by Alexander Fleming in 1929.

Given that there is variation among offspring, and that not all can survive, it is obvious that those most likely to survive are the organisms that have some advantage over their peers. This advantage is local in the sense that it depends on the environment in which the organism is living at the time. Different conditions might favor different individuals in the group. But those who are most likely to survive long enough to reproduce are those individuals who happen to be better suited to their environment. Evolution is not concerned with those individuals that survive but do not reproduce. Not reproducing is the same as not surviving as far as evolution is concerned.

Evolution is a winnowing process. Each generation will have more individuals with favorable characteristics, and fewer with unfavorable characteristics, so there will be gradual change in the group. After many generations this can lead to a change in species. The following example shows how this is possible.

## A Herd of Goats

Here is an example to illustrate evolution. Keep these two facts in mind as this example unfolds: (1) There is variation among offspring, and (2) organisms produce more offspring than can possibly survive to reproduce.

Suppose a herd of goats living in a valley is driven north into the mountains. Perhaps the valley floods, or perhaps too many humans move into the valley. As the goats move north they encounter a hill that creates a fork in their path. Some go to the left, some go to the right, and the herd splits into two parts. As they move further into the mountains the left branch finds themselves on the Western slopes where it rains frequently and the food is plentiful. The Eastern slopes have little rain and sparse food. However, being on the Eastern slopes is not all bad because there are few predators. Most of the wolves and mountain lions are on the Western side where the food is plentiful.

This creates two sets of conditions. It pays for the Western goats to be big, strong, and fast because their main concern is the predators. On the other hand, the Eastern goats should be small and sturdy, small so they require less food, and sturdy to withstand the rigors of starvation during long periods of drought.

Think about the two facts that determine evolution. Variation in offspring will produce goats with a variety of characteristics on both sides of the mountain, ranging from big, fast, strong, and sturdy to small, slow, weak, and frail. The Western goats that are big, fast, and strong are more likely to survive because they have an advantage in escaping from predators. The Eastern goats that are small and sturdy are more likely to survive because big goats may not find enough food to sustain health. Without enough to eat, a large goat will become sick and diseased.

The next generation will take on more of the desirable characteristics for each side because their parents were better able to survive. This will continue from generation to generation until most goats on both sides of the mountain are better adapted to their environment. The Western goats will be noticeably bigger and faster than their small sturdy counterparts on the Eastern slopes of the mountain.

If the two herds are kept separate for a long period, there will come a time when an Eastern goat and a Western goat are unable to produce offspring between them. When this happens, they are said to be distinct species.

It is important to note that there are three species involved in this story; the original herd (we'll call this Species A), the later Western herd (Species B), and the later Eastern herd (Species C). All three are distinct species because the genes of both later herds will be different from the genes in the original herd. The original herd did not "die out." Instead, it evolved into two later herds.

This same process occurs in all species. Any change in the environment produces different conditions that lead to changes in the plants and animals living there. This change is gradual compared to our lifetime, but it can be rapid in geological time.

There is no sudden appearance of a new species and no missing link in this story. The concept of a "missing link" is an illusion. There is no such thing. Each kid is the same species as its parents, and the offspring of that kid will again be of the same species. Species B and C are different from the parent species A, but the change from A to B, or from A to C, is so gradual that no one generation can be called a missing link. From generation to generation the western goats will be noticeably bigger and faster, the eastern goats will be smaller and sturdier, but the change will be so gradual that it will take hundreds of thousands or millions of years before there is a new species.

Darwin knew that his theory required a long period for species to diverge. At the time, the best estimates for the age of the earth were on the order of millions of years, a period that was much too short.

Aristotle thought the earth had existed forever. In 1654 Archbishop James Ussher of Ireland (1581–1656) proposed 4004 B.C. as the date of creation. He even gave the exact date, Sunday, October 23, 4004 B.C. Now we know the earth is about 4.5 billion years old, sufficient time for evolution to work its magic.

Darwin also knew that the mechanism by which creatures evolved was unknown. Now we understand the role of DNA and how it controls the process of life.

## The Role of DNA

In 1944 three biochemists working at the Rockefeller Institute, Oswald T. Avry, Colin M. MacLeod, and Maclyn McCarty, were able to show that nucleic acid was the substance of inheritance. Then Watson and Crick discovered the double helix structure of DNA in 1953, and this led to a flurry of activity to discover more and more of the details that describe how life works. Now we know that life works very much like a digital computer. Computers are controlled in a manner that is eerily similar to the way DNA controls the functions of life. Consider the following analogies:

- Digital computers use a binary code, 1 and 0. DNA uses a quaternary code consisting of the four nucleotides adenine, thymine, guanine, and cytosine. They are symbolized by the letters A, T, G, C. These codes are used by both the computer and the cell over and over in a step-by-step procedure with a few variations that produce amazingly diverse results.

- When a computer performs a certain task, whether it is dialing a phone, or perhaps recalling a picture from memory, it is controlled by instructions stored in its own memory. Likewise, when a certain task is performed in a living cell, instructions are followed from code stored in the cell's DNA.

- In the computer an instruction is fetched from memory and copied into the central processing unit (CPU). In a cell an instruction is transported from the DNA to a ribosome by messenger RNA. In both computer and cell the instruction is only copied, not moved. The instruction remains undisturbed in computer memory, and it remains undisturbed in DNA.

- Once in the CPU the instruction code word causes one of a small number of possible operations to be performed. Once in the ribosome the RNA code word causes one of a small number of possible operations to be performed. In each case, it takes many similar operations to accomplish one overall task.

These are but a few of the similarities between the intricate ways computers and cells work. This similarity extends to the details of how code-words control all operations. The above discussion explains *what* a computer or cell does. It does not explain *how*. A complete explanation of how a computer or cell works is beyond our scope, but the following cursory discussion should give you some idea.

A computer does not store nor operate on numbers such as 1 or 0. It is an electronic device that operates with voltages and currents. If we let +4V represent "1" and +1V represent "0," then these voltages can control the current through electronic valves (transistors). Therefore, the binary code symbols that we think of as (0, 1) are, in reality, voltages (+1V, +4V). The transistor bias voltage can be set so that a +4V input allows current to flow through the device, and a +1V input inhibits current flow. This gives us a method of converting the binary code (0, 1) directly into electronic operations, and thus allows control of the computer operations.

In the early computers there were 16 different operations that the CPU could perform. There are 16 different combinations of four-bit sequences, from 0000, 0001, 0010, 0011, etc., to 1111.

Therefore the operations code, "op code," consisted of four bits. These sixteen operations were the total sum of all possible operations in early computers.

In a similar manner, three-letter combinations of the molecules (A, T, C, G) such as CTG, TAC, and AGC generate amino acids that control the production of proteins in the cell. Twenty amino acids are used in this process. Three letters are necessary to specify 20 amino acids.

Think of it this way: Four different symbols, A, T, C, G could specify four amino acids. Two symbols at a time, such as AG or GC, give 16 possible combinations. These two-symbol combinations could represent 16 different amino acids, still not enough. With three-letter combinations, such as AGT or GCA, there are 64 different combinations, too many. But that is OK, so the op code for cells consists of three letters.

In place of voltages and currents, the cell operates chemically by producing proteins that are sequences of amino acids. These 20 amino acids are strung together in thousands of unique combinations to construct proteins, all controlled by 3-tuples of DNA code.

This cursory explanation should give you some idea of the similarities between computers and cells. As we dig deeper into the details of exactly how computers and cells accomplish their operation this similarity only increases. This is both remarkable and somehow disturbing on a number of fronts.

Ever since the public became generally aware of computers (generally between 1960 and 1980) the question has been asked, "Could computers become so intelligent that they will become our masters?" The troubling answer is "yes, they can."

In 1965 Gordon Moore, who was director of R&D at Fairchild Semiconductor, published an article in which he predicted that computing capacity would double every year. He later revised his prediction to once every two years, and this has held true until today Intel's 8080 microprocessor containing 4,500 transistors was released in 1974. Today the highest density chips available contain 4.5 billion transistors. To put this in perspective here are these numbers written out.

1974          4,500 transistors

Today         4,500,000,000 transistors

This is a one million-fold increase in 40 years, which agrees with Moore's Law.

We have reached the limit of Moore's Law, but don't let that reassure you. Engineers and scientists are now working on quantum computing, and if perfected this will give a great increase in computing capacity in one fell swoop. This can only increase the likelihood that computers can someday match our intelligence.

With that scary note we must leave the comparison between computers and life here. Let us return to our main theme, the way evolution works.

## Gregor Mendel

An intermediate step in understanding evolution is the rules of inheritance. This knowledge was supplied a few years after the publication of Darwin's and Wallace's theory by a reclusive monk named Gregor Mendel (1822–1884).

If evolution is really the way things happen, then why isn't everyone the same? We all have two parents, four grandparents, eight great-grandparents, and so on. With all those diverse genes in us, we should be a duke's mixture of all possible characteristics. The inheritance of characteristics should be like mixing paint, or so most people thought before Mendel's work became known. Mixing red and blue paint gives purple paint. Why doesn't inheritance work like that? Everyone should display average characteristics, and we should all look alike, talk alike, have the same abilities, and have the same interests. But this does not happen. Diversity among offspring is the norm, and this diversity never seems to diminish.

There must be some mechanism that keeps all members of a species from becoming average, and instead maintains diversity. This mechanism works by rules that were published by Gregor Mendel in the late 1860's. The understanding of the actual mechanism itself would have to wait until the discovery of DNA.

Diversity of organisms is necessary for evolution to work. Recall that one of the guiding principles of Darwin's theory is variation among offspring. If there is no variation then nature cannot differentiate, it cannot favor one individual over another. Those that survive would be chosen at random. But nature does not work that way. Although the choice has some randomness in it, the overall average favors those individuals with the "correct" set of characteristics.

When Alfred Russell Wallace and Charles Darwin published their theory of evolution in 1858 there was no known mechanism to carry variations over from generation to generation. If mating occurred between creatures that varied greatly, then why would not these variations merge into the average? How could distinct characteristics be passed on from generation to generation without disappearing?

The answer came in the form of two obscure papers published in 1865 and 1869 by Gregor Mendel, an Augustinian Monk in Brunn, Austria. His papers were published in the *Transactions of the Brunn Natural History Society*, an obscure publication not read by many. Also, his papers contained both botany and math, so botanists ignored his paper because of the math, and mathematically inclined scientists ignored his paper because of the botany. Thus his paper was largely ignored until 1900 when Hugo de Vries came across Mendel's papers after formulating his own theory, which agreed with Mendel's theory.

Mendel did his work over many years after analyzing literally thousands of pea plants. He crossed varieties of these plants in a garden at the monastery to determine the results of combining seven characteristics of the plants:

1. Flower color:       Purple or White
2. Flower position:    Axil or Terminal
3. Stem length         Long or Short
4. Seed shape          Round or Wrinkled
5. Seed color          Yellow or Green
6. Pod shape           Inflated or Constricted
7. Pod color           Yellow or Green

Common pea plants are ideal for this experiment because they can be grown easily in large numbers and their reproduction can be controlled. Their flowers have both male and female organs, so they can self-pollinate or cross-pollinate. Also, Mendel was fortunate to pick seven characteristics that are mutually independent. This means that the stem length does not influence the flower color, for example, or the seed shape does not influence the pod shape. Each of these seven characteristics have minimum influence on the others.

This is not so for other characteristics. For example, height in humans depends on the size of several different body parts—lower leg, upper leg, torso, neck, and head. A different gene controls each size, so overall height is the culmination of these various sizes. Skin color is another characteristic that is controlled by many genes. And so is eye color.

Here is a description of one experiment that should illustrate Mendel's genius. This experiment begins with two pea plants that are purebred for seed color, one yellow, and one green. When cross-pollinated they produce only yellow seeds.

The plants that grow from these yellow seeds are then self-pollinated, and they produce yellow and green seeds in a 3:1 ratio. There are three yellow seeds for every green seed. Or, to put it another way, three-fourths of the seeds are yellow and one-fourth are green.

The plants that grow from these seeds are again self-pollinated, and here is where it gets really confusing. The yellow seeds produce yellow to green seeds in a 5:1 ratio, and the green seeds produce only green seeds. Now I ask you, how could anyone but a genius make any sense out of all this?

After several years of recording results such as these for all seven characteristics he determined five principles that establish the laws of inheritance. It is an amazing fact that the five principles that he established from these experiments apply equally to all living things. These are:

1. A specific "heredity factor" controls each physical characteristic of a living organism. Mendel envisioned this heredity factor as some kind of particle. We now call it a gene.

2. These heredity factors exist in pairs. Although there may be several colors for eyes, each person carries exactly two of these.

3. Each offspring receives one factor from the father and one from the mother, thus maintaining exactly two factors in the offspring.

4. There is an equal probability that either one of the father's factors and either one of the mother's factors will be inherited by the offspring.

5. Some factors are dominant and some are recessive. Offspring display the dominant factors while carrying both.

Now I want to give you an example, but first we need some technical terms. A *gene* is defined as a section of DNA. This section must be long enough to have sufficient copying fidelity to serve as a viable unit of natural selection. The particular content of this gene is called an *allele*.

The same gene in different people may contain different alleles. The hair on one person may be brown, another black, and another blond, so these three people have different alleles for hair color. But in every case each person carries exactly two alleles in that one gene. The dominant allele is the one that determines hair color.

Here is an example to demonstrate the application of Mendel's five principles. Suppose a particular goat can have one of four colors, black, gray, tan, and white. Suppose the mother carries black and tan alleles, and the father carries tan and white alleles.

Since an offspring will carry one of the mother alleles and one of the father alleles, this gives the picture in Fig. 4.1. This chart is called a Punnett square, named after the early 20<sup>th</sup> century

English geneticist Reginald Punnett. To construct a Punnett square, start by drawing two vertical and two horizontal lines (a tic-tac-toe diagram.) Place the alleles for one goat across the top; say black (B) and tan (T). Down the left side list the other goats alleles, T and W, as shown in the left diagram. Then fill in the blanks with the possible offspring pairs shown in the right diagram.

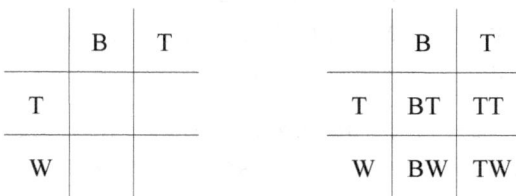

**Figure 4.1. Punnett diagram for goat colors.**

Each of these possible offspring combinations is equally likely to occur, meaning that about ¼ of all offspring from these two parents will have black/tan, ¼ will have black/white, ¼ will have tan/tan, and ¼ will have tan/white alleles.

Now suppose that the order of dominance is black, gray, tan, and white, in that order. Thus, the color of a black/tan offspring or a black/white offspring will be black. The color of a tan/tan or tan/white offspring will be tan. In this example the only possible colors for the offspring are black and tan, and these occur with equal probability.

White is last in the order of dominance. The only way for a goat to be white is for it to inherit two white alleles, one from each parent. This is possible if the two parents carry gray/white and tan/white, for example. In that case, about ¼ of the offspring will be white, ¼ will be tan, and ½ will be gray. (Remember that gray is dominant over both tan and white.) You can draw your own Punnett diagram to see that this is correct.

It is amazing that Gregor Mendel could have puzzled this all out by studying pea plants. The fact that each plant carries two alleles with one of them being passed on to each offspring, thus assuring that each plant contains only two alleles for a particular trait, must have caused many sleepless nights with much thought. For him to discover this, plus the fact that alleles are ordered for dominance, is truly amazing, especially since no mechanism for this had yet been discovered. Now the mechanism for this is known through the discovery of DNA, the double helix structure, and the subsequent identification of genes with sequences of this DNA.

We are now in a position to solve Mendel's riddle of the seeds. He began with purebreds, meaning that the alleles in each plant were the same, YY for one plant and GG for the other. When these were cross-pollinated the first generation had identical alleles in all plants, as shown in Fig. 4.2. The yellow seed color s dominant, so this explains why all seeds in the first generation were yellow.

|   | G | G |
|---|----|----|
| Y | YG | YG |
| Y | YG | YG |

|   | Y | G |
|---|----|----|
| Y | YY | YG |
| G | YG | GG |

**Fig. 4.2. First Generation**     **Fig. 4.3. Second Generation**

After the first generation, self-pollination is the rule. For the second generation each self-pollination results in the four possible allele combinations in Fig. 4.3. Since yellow is

22

dominant, there will be three yellow seeds for every green seed, giving the ratio 3:1. Three fourths of the seeds will be yellow and one fourth will be green.

Mendel grew several plants in each generation. Figures 4.2 and 4.3 simplify things down to the bare essentials, making the task of interpreting the results much easier than Mendel had it. Now for the final step each of these many plants is again self-pollinated. Figure 4.4 shows the simplified results, where there is one Punnett diagram for each outcome in Fig. 4.3.

|   | Y  | Y  |   | Y  | G  |   | Y  | G  |   | G  | G  |
|---|----|----|---|----|----|---|----|----|---|----|----|
| Y | YY | YY | Y | YY | YG | Y | YY | YG | G | GG | GG |
| Y | YY | YY | G | YG | GG | G | YG | GG | G | GG | GG |

**Fig. 4.4. Third Generation**

If we count yellow and green seeds in this generation there are four yellow seeds in the left Punnett diagram, three yellow and one green in the second diagram, three yellow and one green in the third diagram, and four green seeds in the last diagram. The yellow-seed parents in the first three diagrams produced ten yellow seeds to two green seeds, or a 5:1 ratio, and the green-seed parents produced only green seeds. This agrees with Mendel's results.

It all fits together to form the most important discovery in science. The fossil evidence, Darwin's and Wallace's theory, the way DNA controls the process of life, and Mendel's rules of

23

inheritance all converge to form a complete description of evolution.

## The Human Story

Where did we humans come from, and how did we get here? In exactly the same way the goat herd in the above example evolved into two later species.

Our nearest cousins are the Chimpanzees and Bonobos. A Bonobo is a midget chimp, and although they look alike except for size, they are distinct species. Chimps quarrel, bond, murder, deceive, and make tools. It is but a small exaggeration to say that Bonobos live the ideal life, making love all day. The Bonobo split from Chimps about 2 million years ago, so I will refer to both as Chimps.

Humans and Chimps evolved from an earlier species in the same manner that Species B and C evolved from the original goat Species. Humans and Chimps encountered the equivalent of the hill that split the goat herd into two parts sometime before six million years ago.

No one knows for sure just what the "hill" was that split our lineage from the chimp lineage. It seems likely that the split occurred when some of our ancestors lived in an area with few trees, while others ived in the forest. We are descendants of those on the plains, while chimps descended from the tree dwellers.

Some of those living in the grassland were better able to walk upright than others. This gave the upright walkers an advantage in surviving, or perhaps an advantage in finding a mate. No one knows why walking upright developed, but it must have been an advantage that allowed our ancestors to survive.

One suggestion is that standing upright made it easier to see over tall grass to avoid predators. Another is that standing upright exposes sexual genitalia, thus increasing the likelihood of attracting a mate.

The faster you can run the easier it is to escape. So why did we not develop the ability to run on four legs? Almost any four legged animal can outrun a human.

My suggestion, which I have not seen in print, is a case of finding the best route to safety. There is a universal principle that says you must start from where you are. You cannot start a journey or process from some other place or some other condition. The chimps living on the plain were able to climb, but were unable to run effectively, so they had to make some changes in order to survive.

If chimps ran on four legs they would lose the ability to climb trees, and climbing trees is an obvious advantage in escaping predators. Running on two legs permits some gain in speed while retaining the ability to climb. Therefore more upright walking hominids survived and passed on their genes to their descendants, who in turn had an advantage over those who did not walk upright. After many generations all hominids walked upright.

In any event, the first upright walking ape appeared on the African Savana about 6 million years ago. Although this creature was our ancestor, it was more apelike than human. Scientists label all upright walking creatures "hominid," so this was the first hominid.

This initial hominid spawned at least 19 different species that lived at one time or another during the last 6 million years. Why we are the last and only surviving hominid is another mystery. But here we are, overpopulating the world and destroying the environment in an effort to hasten our own demise.

## Culture

The fossil record, combined with other archaeological discoveries, indicate that culture developed sometime after fifty thousand years ago. After that time some of our ancestors were buried with ceremony. Flowers and trinkets are found in graves, with the corpse sitting in an upright position. Red ochre, found in many graves, must have been a substitute for blood. This, along with food, were provisions for the journey ahead.

Seashells with holes drilled in them made bracelets and necklaces. Carved statuettes and engraved stone date back to around forty thousand years ago. But the most spectacular evidence for early culture is found in the cave paintings that have survived for 35 to 40-thousand years.

Why did culture gradually develop after 50 thousand years ago? Probably because speech developed at that time. We don't know when speech developed, but it's a good bet that speech developed before culture. It seems likely that culture followed closely on the heels of the ability to communicate ideas.

On the other hand, speech may have developed much earlier. The evidence for this is the presence of Broca's area in the skull of the famous Turkana Boy, an almost complete skeleton of a 1.6 million year old *Homo Ergaster* found by Richard Leaky and Alan Walker in 1984. Broca's area is a protrusion on the left frontal lobe of the cortex. This lump leaves an indention in the inside of the skull.

Paul Broca was a nineteenth-century French physician who noticed that injuries to this part of the brain made speaking difficult. Thus Broca's area is involved with the production of speech. But it also is involved with memory and executive functions not related to speech. Thus the presence of Broca's area does not guarantee that the Turkana Boy was able to speak. We simply don't know when speech developed, but the bulk of scant evidence favors 50 thousand years ago.

It is meaningless to put a date on the development of our own species, homo sapiens, because evolution is an ongoing process. But we know from fossilized skeletons that our ancestors looked very much like us as early as 200,000 years ago. You will see statements in the literature that date the beginning of homo sapiens anytime between 250,000 to 150,000 years ago.

## A Poem from the BC Comic Strip

There once was a preacher named Charley D.
whose degree was in Christianity
But he chose to take up botany
so he sailed aboard the HMS B.
A fox of a man, on a dog of a ship,
allowed his Christianity to slip
and concocted a theory while at sea,
designed to make monkeys of you and me.

## Books by Dwight F. Mix
Technical LAP Series:

Volume 1 in the Technical LAP Series. A description of mathematical concepts in plain language. Vector spaces, symbolic derivative and steepest descent, matrix of transformation, least squares and the pseudo inverse, probability and random variables, eigenvectors, crisp and fuzzy logic, and entropy. These concepts are first defined and then illustrated with engineering applications to add context.

LEARNING ACTIVITY PACKAGES
FOR PATTERN RECOGNITION

*Dwight F. Mix*

Volume 2 in the Technical LAP Series. What is entropy? What is a fuzzy system? How can one use these in pattern recognition? What can artificial neural networks do? How can one reduce dimensions with a minimum loss of information, or design templates that take into account the similarities and differences between classes?

CONTINUOUS-TIME TRANSFORMS FOR ENGINEERS AND SCIENTISTS

*Dwight F Mix*

Volume 3 in the Technical LAP Series. Aimed at students, practicing engineers, and scientists, this text presents unique insight into the continuous-time Fourier transform and series. How to plot complex exponential signals. How to use this technique to calculate the transform at a specific frequency by finding the center of mass in the complex plane. Three chapters (LAPS) are devoted to the Laplace transform, one on the forward transform, one on the inverse transform, and one on properties.

Student's Guide to Discrete
Fourier and z Transforms, Sampling,
Multirate Processing, and the FFT

*Dwight F. Mix*

Volume 4 in the Technical LAP Series. For students, practicing engineers, and scientists. Discrete-time Fourier transforms, z transforms, sampling, multi-rate processing, and the fast Fourier transform. An introduction to the pulse sorting transform. This is a unique modification to the Fourier transform for the purpose of sorting pulse signals.

# Student's Guide to Fourier, Laplace, and z Transcorms

## Technical LAP Series, Vol. 5

## Dwight F. Mix

Volume 5 in the Technical LAP Series combines Volumes 3 and 4 into one volume. After showing how to plot complex exponential signals, this text introduces all four forms of the Fourier transform. The forward and inverse Laplace transform for continuous-time signals, and the forward and inverse z transform for discrete-time signals. Insight into the process of finding transforms. Specifically how to estimate the Fourier transform of both continuous-time and discrete-time signals from Argand plots of complex exponential signals.

Student's
Guide to Analog
Systems.

State Equations, Transforms,
Convolution, Controllability
and Observability

Dwight F Mix

The three methods for finding the response of an LTI system are differential equations, transform methods, and convolution. Under the conditions of controllability and observability these three methods are equivalent. If the system is not LTI there are no general methods for finding the response. This text introduces the LTI properties plus controllability and observability, and shows their connection to all three methods.

*Filter Design Techniques*

Analog and Digital Filters. Frequency Selective and Matched
Filters. Adaptive Matched Filter and Template Design

Dwight F. Mix

Filter Design Techniques explains how to design analog, digital, and matched filters. It is intended for practicing engineers and scientists who have a background in Fourier, Laplace, and z transforms.

Part 1 is concerned with analog Butterworth and Chebyshev filter design. Part 2 explains IIR and FIR digital filter design. Part 3 introduces adaptive design of templates for pattern recognition and matched filters for signal detection.

The design technique in Part 3 takes into account both the signal to be detected and the differences and similarities to all the other signals or patterns of interest.

34